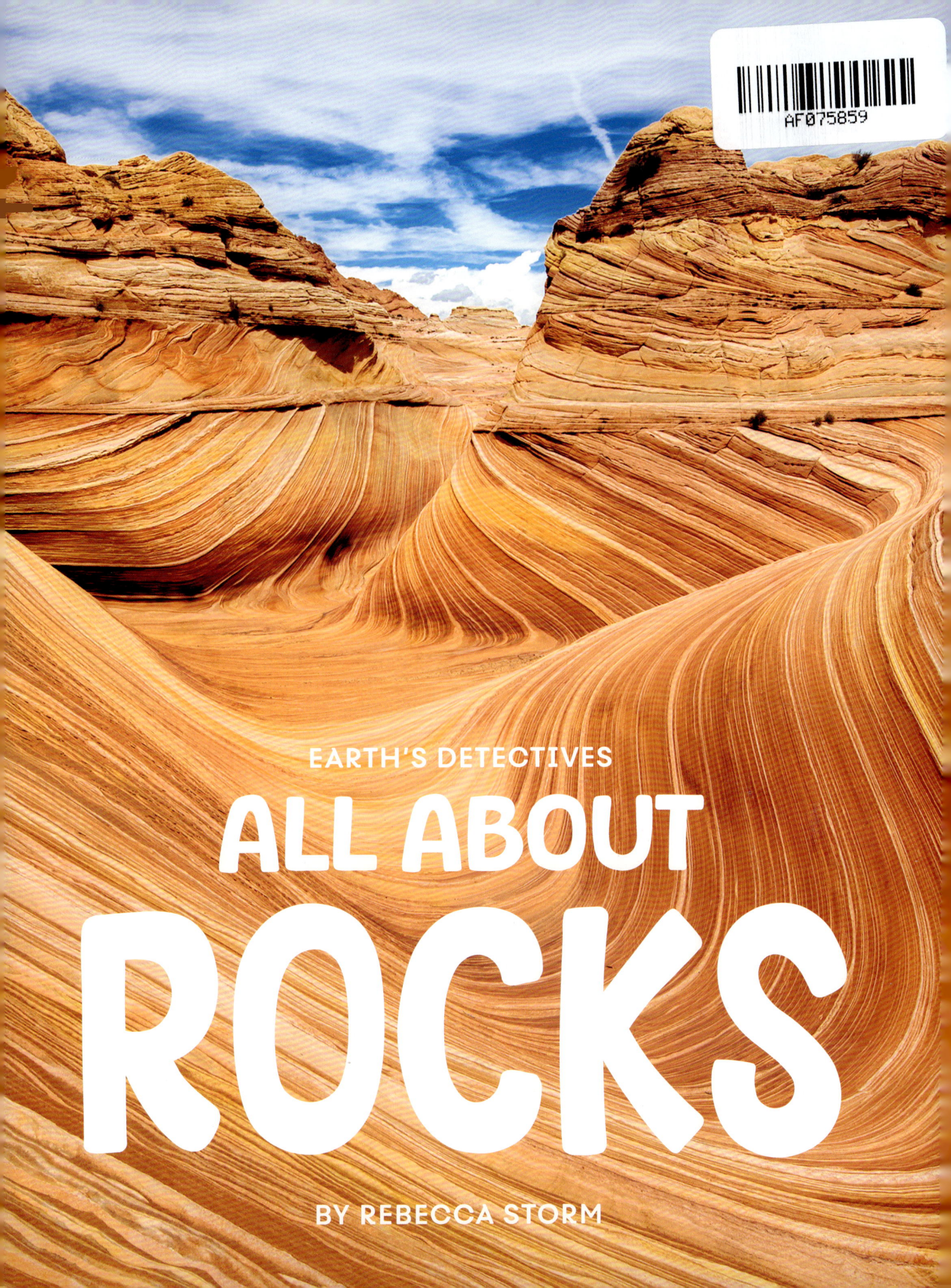

EARTH'S DETECTIVES

ALL ABOUT ROCKS

BY REBECCA STORM

CONTENTS

WHAT ARE ROCKS?	4	HOW TO FIND ROCKS	20
HOW DO ROCKS FORM?	6	IGNEOUS ROCKS	22
IGNEOUS ROCK: GRANITE	8	SEDIMENTARY ROCKS	24
SEDIMENTARY ROCK: SANDSTONE	10	METAMORPHIC ROCKS	26
METAMORPHIC ROCK: GNEISS	12	RECORD-BREAKING ROCKS	28
WHAT CAN WE LEARN FROM ROCKS?	14	TRUE OR FALSE?	30
ROCK DETECTIVES	16	GLOSSARY & INDEX	32
AMAZING ROCKY PLACES	18		

Words in **BOLD** can be found in the glossary.

Copyright © 2025 Hungry Tomato Ltd

First published in 2025 by Hungry Tomato Ltd
F15, Old Bakery Studios, Blewetts Wharf, Malpas Road, Truro, Cornwall, TR1 1QH, UK.

No part of this publication may be reproduced, stored in a retrieval system, or transmitted in any form or by any means, electronic, mechanical, photocopying, recording, or otherwise, without prior written permission of the copyright owner.

A CIP catalogue record for this book is available from the British Library.

ISBN 9781835690802

Printed in China

Discover more at
www.hungrytomato.com

Picture credits:
Abbreviations: m-middle, t-top, l-left, r-right, bg-background.

Chris and Helen Pellant: 9m, 13t. All Canada Photos/ Alamy: 13b. Shutterstock: 9b; 3DML 20b (rock pick); Aleksandr Pobedimskiy 25tl; Alones 30bl; Arctic ice 4b; ArganaHf 24bl; Artsiom P 30b; Artur Vaisman 14br; Barou abdennaser 10tr; Bill Florence 31ml; Bjoern Wylezich 23tl; CardIrin 17m; Daniele Collova 5br; daulon 8b; EmilEn4ev 15tl; Engineer studio 15mr; Evgeny Haritonov 27mr; Ground Picture 21tl; iadams 20b(magnifying glass); IamTK 26bl; Iva Vagnerova 21br; Jannarong 31tr; Kazuki Yamakawa 5m; Kevin Eng FC; KrimKate 28b (peridotite); LordLukhan FCbg; Matteo Chinellato 29ml; Mark Brandon 11br; Michael Kaercher 18t; michal812 28bm (gabro); Mr.Navapruet Promthong 24mr; mykolastock 20b(backpack, camera); Paulo Miguel Costa 2-3bg, 19t; PeopleImages.com - Yuri A 21ml; Pornpimon Ainkaew 16m; Pung 1bg, 11t; PurMoon 31ml (black sand); Stas Malyarevsky 28br (pumice); SeventyFour 21mr; Stevo Elliott 29tl; Tania Stout 19b; TR_Studio 30tr; Trismegist san 29br; Tyler Boyes 22tl; Rainer Lesniewski 15bl; Vadym Lavra 31br; Vasiliy Ptitsyn 14tl; Vixit 18br; vvoe 22bl, 23mr, 24tl, 26mr, 27bl; wawritto 16br; Yes058 Montree Nanta 22mr, 23bl, 25bl, 26tl, 27tl; yothinpi 25mr; zendograph 28m;

Every effort has been made to trace the copyright holders, and we apologise in advance for any unintentional omissions. We would be pleased to insert the appropriate acknowledgements in any subsequent edition of this publication.

WHAT ARE ROCKS?

You may not notice it, but rocks are all around us! This is because Earth's surface is made of rocks. They're even under forests and oceans.

There are many different kinds of rocks. Some are very hard and difficult to break, while others are soft and easy to shape.

Rocks can form around **GIANT MOUNTAINS...**

or **TINY GRAINS THAT MAKE UP SAND!**

Rocks shape all of the **LANDSCAPES AROUND US**. See how many different rocks you can identify next time you are out and about. We use rocks to make things, too. Roads, buildings, and statues are all made of rocks. The plates you eat from and the chalk you draw with are made of rock, too!

Impressive rocky sites, like the Grand Canyon in Arizona, USA, attract millions of tourists every year.

GRAND CANYON, USA

It is incredible how the Earth has made such beautiful things out of rock, like **Wave Rock in Australia.**

HOW DO ROCKS FORM?

Rocks are forming and changing all the time. This process is called the "rock cycle". It happens on the surface of the Earth and underground, where it's very hot, and it takes millions of years.

Hot, molten rock flows from volcanoes. It cools to form igneous rock.

Underground, igneous and sedimentary rocks are heated and squashed, which turns them into metamorphic rock.

Above ground, hot and cold weather, wind and moving water wear rocks down into tiny pieces called "sediment".

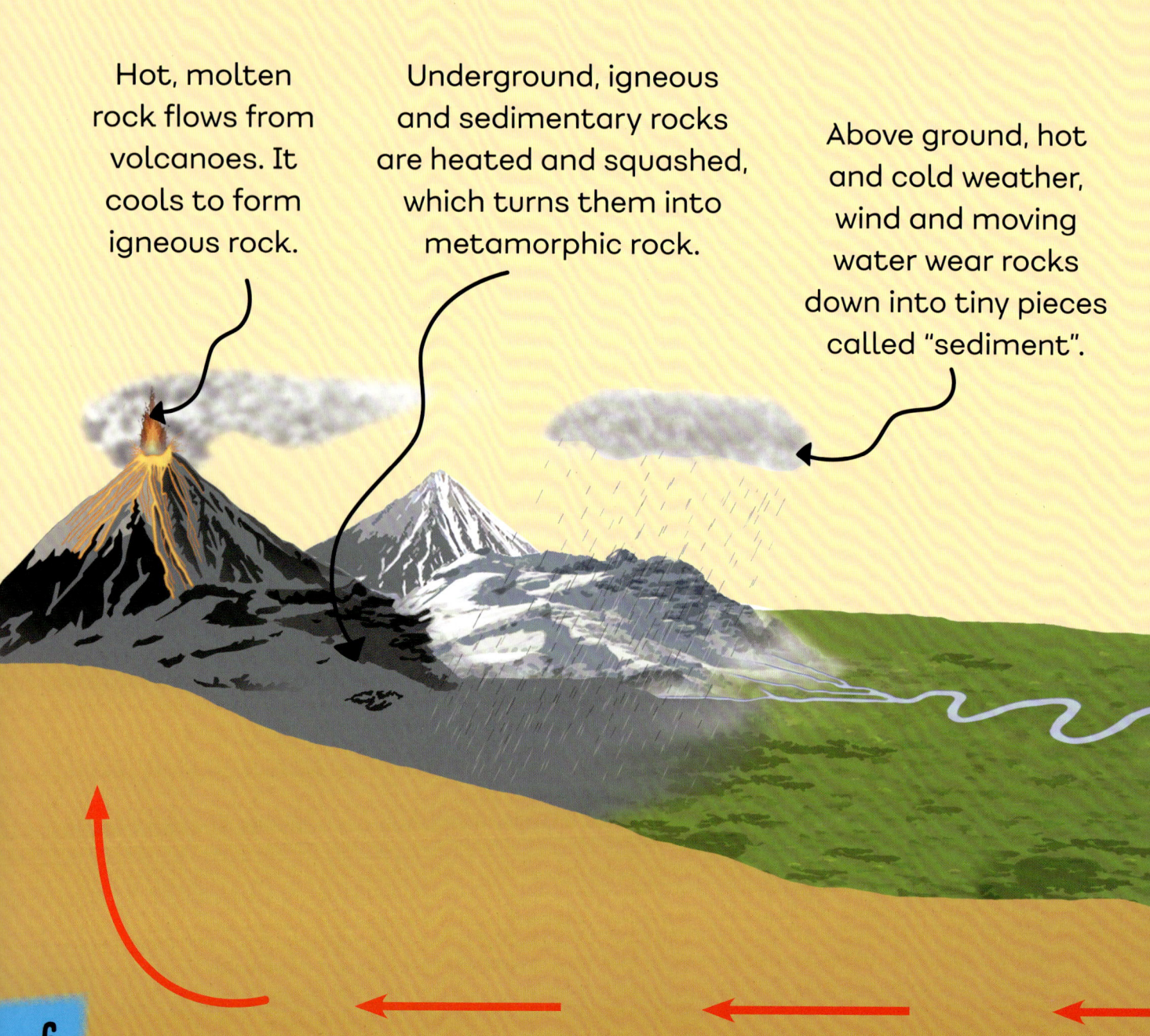

This is how the world's three types of rock form:

- **Igneous rock** forms when **molten** rock cools down and hardens.

- **Sedimentary rock** is formed when small, worn off pieces of other rocks become joined together in layers.

- **Metamorphic rock** forms when heat or pressure causes rocks to change their structure or mineral **composition**.

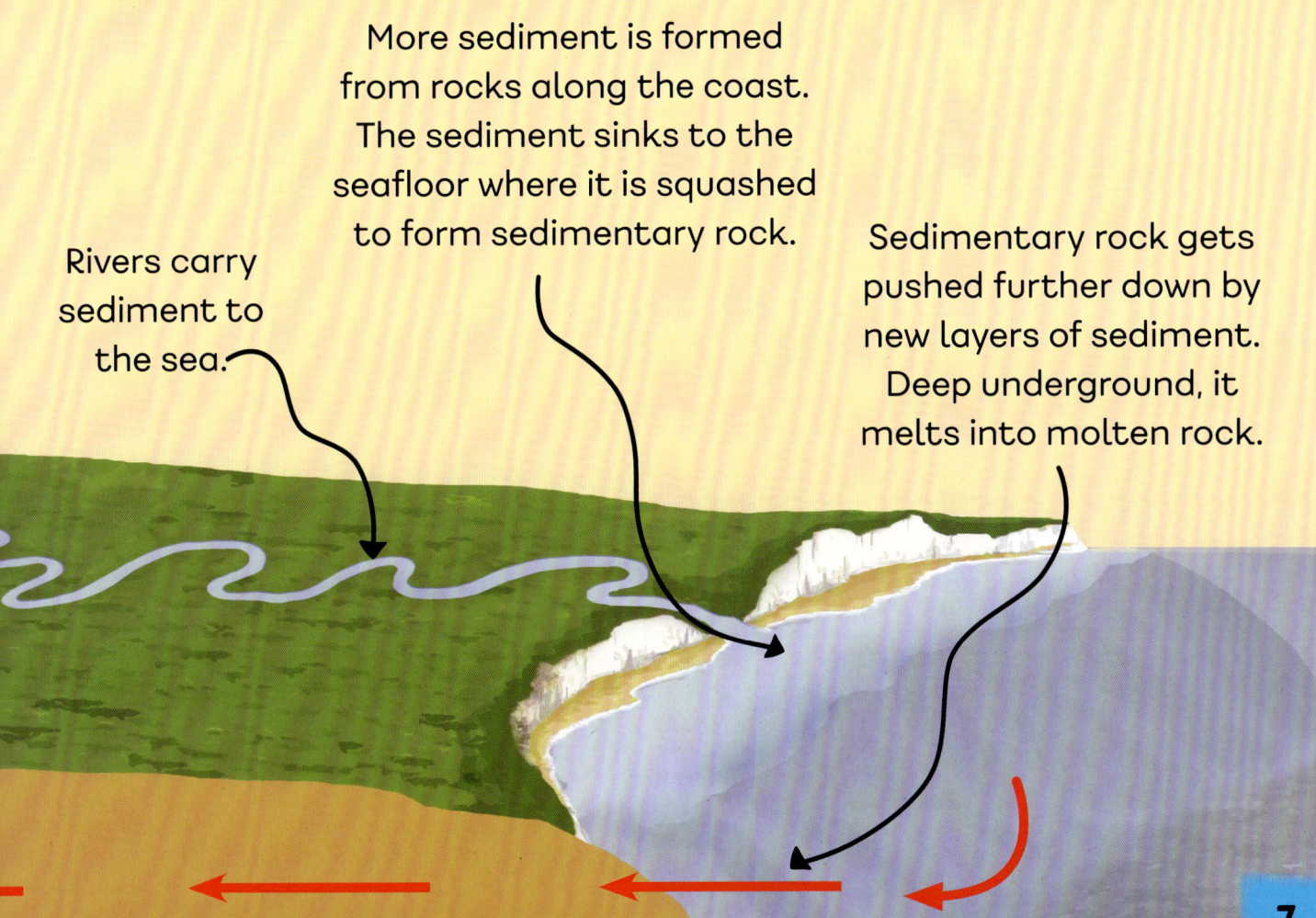

More sediment is formed from rocks along the coast. The sediment sinks to the seafloor where it is squashed to form sedimentary rock.

Rivers carry sediment to the sea.

Sedimentary rock gets pushed further down by new layers of sediment. Deep underground, it melts into molten rock.

IGNEOUS ROCK: GRANITE

Granite is one example of an igneous rock. It is formed of hot, molten rock from deep under the Earth's surface.

Molten rock has a few names. When it's underground, it's called **MAGMA**. When it's reached Earth's surface, like when it erupts out of a volcano, it's called **LAVA**. Once it's cooled and hardened, which can happen above or underground, it's called **IGNEOUS ROCK.**

Granite is formed from rock that cools down inside the Earth.

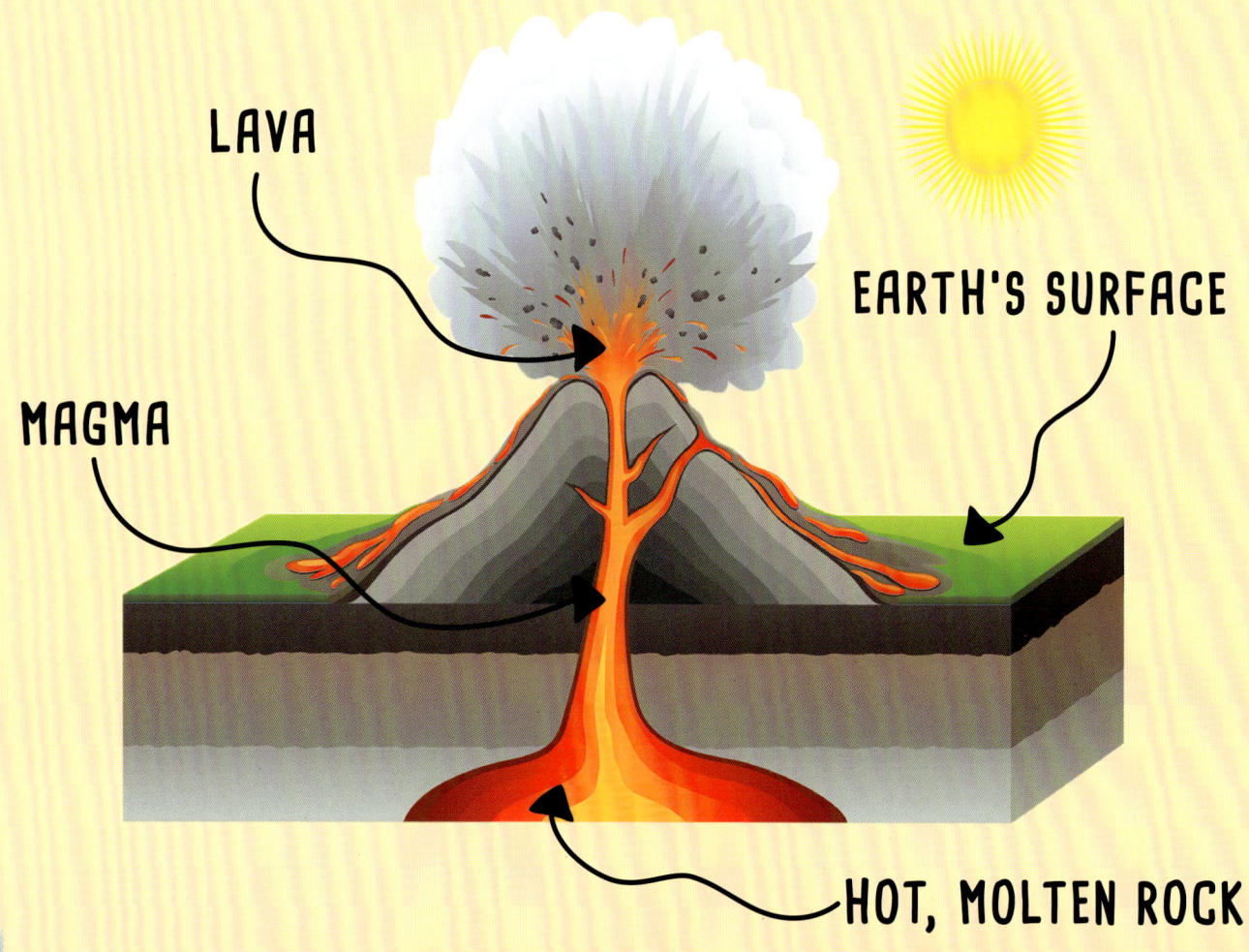

Like all rocks, granite is made from **minerals** - tiny pieces of material that join together to form rocks. Sometimes, they form as **crystals**. There are thousands of minerals in the world, and they can look very different.

Granite is made of three different minerals: quartz, mica, and feldspar. They were all fused together when the granite formed.

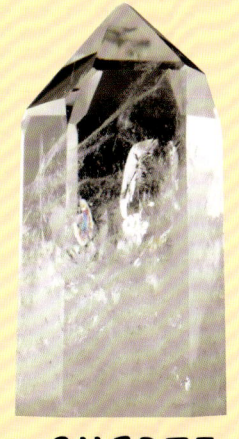

QUARTZ
(white/grey areas)

MICA
(black areas)

FELDSPAR
(pink areas)

SEDIMENTARY ROCK: SANDSTONE

Sandstone is an example of sedimentary rock. It is made of layers of sand that have been pushed together to make a solid block of rock.

When you look at sand through a **microscope**, you can see the individual **grains**. The grains are tiny pieces of a mineral called quartz, mixed with animal shells or other minerals.

SAND UNDER A MICROSCOPE

Ice, heat, wind, and moving water wear grains of sand off rocks which pile up in layers.

As more layers form, the layers at the bottom are squeezed together to form sandstone. This can happen in a desert, river, sea, or ocean.

In the sandstone of Coyote Buttes, in Arizona, USA, it's easy to see the layers of sand the rock is made of. This rock was formed over millions of years. Its layers are vibrant oranges, yellows, reds, and pinks because of the rich minerals inside.

Almost all **fossils** are found inside sedimentary rock, like sandstone and limestone. The layers of sediment and rock made great conditions for preserving animal and plant remains for millions of years.

METAMORPHIC ROCK: GNEISS

Gneiss (pronounced "nice") is an example of metamorphic rock. Rocks become metamorphic when they change and become a different kind of rock.

Deep under Earth's surface, there is a lot of heat and pressure. When the rocks there are heated and squeezed, the minerals inside them change. This makes them a completely new type of rock!

When you compare a rock before and after it has changed, it can be very different! Look what happens when granite turns into gneiss:

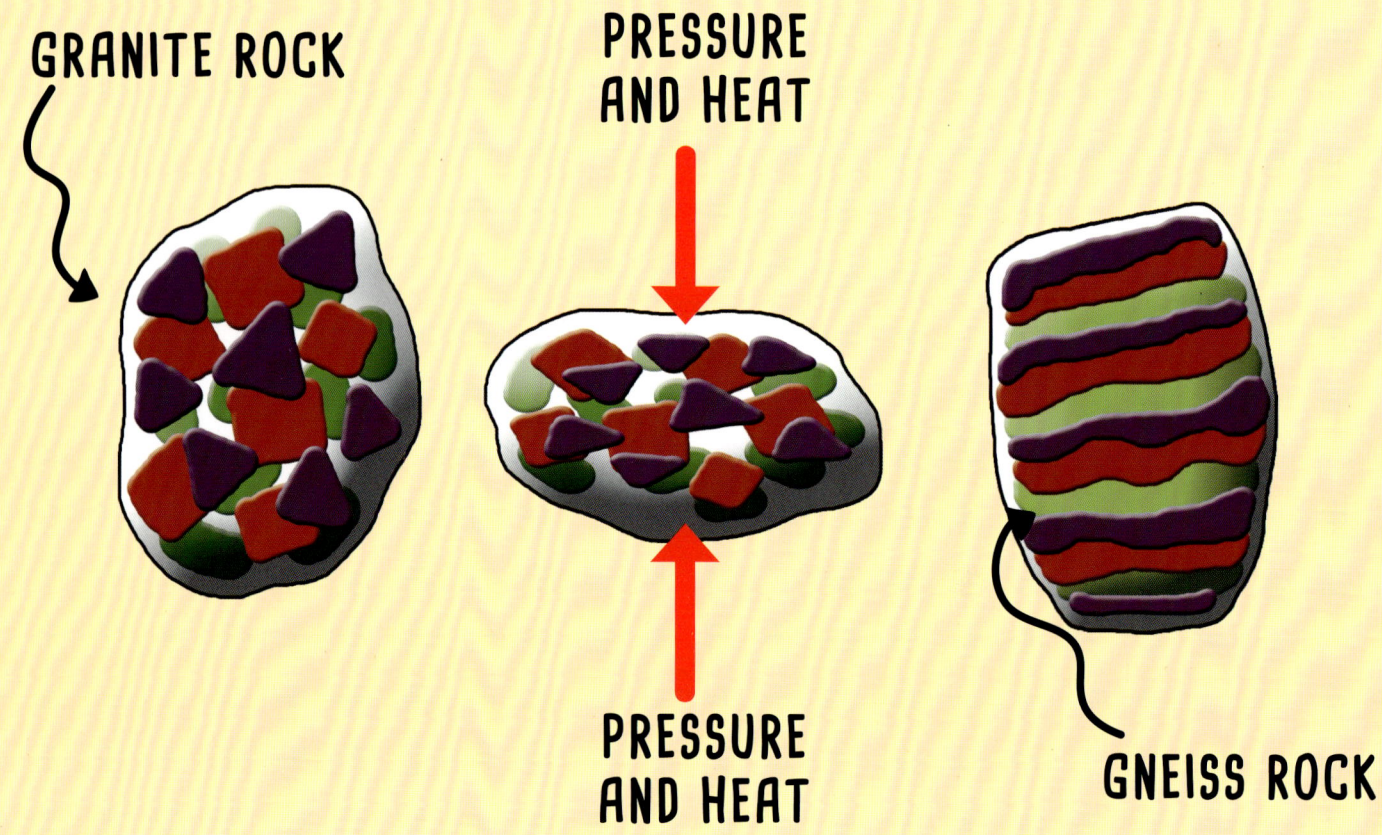

When you compare granite with gneiss, you can see that the pattern has changed. The minerals in granite are speckled. In gneiss, they look stripy!

GRANITE
(Igneous rock)

GNEISS
(Metamorphic rock)

This photograph shows part of the Canadian Shield in northern Canada, an area which is mostly made of gneiss. The region is one of the largest areas of ancient metamorphic rock on Earth – some of the rocks here are more than 4 billion years old.

WHAT CAN WE LEARN FROM ROCKS?

Rocks can tell us all sorts of things about what Earth was like in the past - it's like looking back in time.

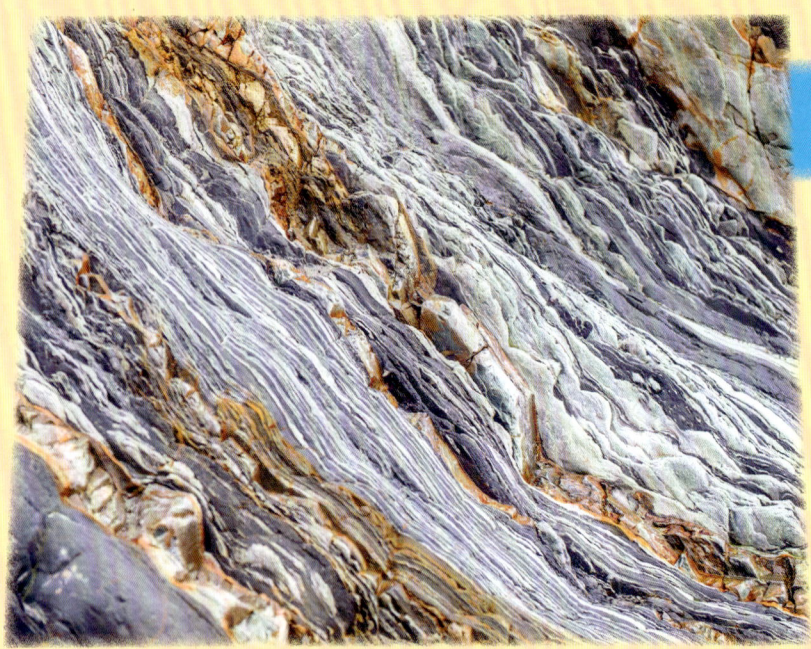

Changing habitats

By studying the different layers that make up rocks, scientists can see when habitats changed, for example, from deserts into wetlands.

Climate change

Rocks teach us about **climate**. Glaciers create sediment called "till". Where till is found, we know the climate was once cold. In hot climates, salt layers form on the ground and get preserved in rock. These clues help us track **climate change**.

Natural disasters

Rocks can hold information about natural disasters too. **Dating** rocks around volcanoes allows scientists to figure out when volcanic eruptions happened and how powerful they were.

Prehistoric life

Some layers of rock contain fossils of ancient plants and animals. This means we can learn about the living things that used to rule the Earth, even if those **species** no longer exists!

Changing lands

Scientists found rocks in America, Europe, and Africa that were identical. From studying their age and composition, scientists have shown that all the land on Earth was once joined together!

ROCK DETECTIVES

There's so much to learn from rocks, but how do we uncover the secrets they hold?

Scientists who study rocks are called **GEOLOGISTS**. It's thanks to their studies that we have learnt so much about rocks and the power and history of planet Earth.

Geologists spend lots of time doing fieldwork. This can include collecting samples of rock using hammers and **chisels**, drilling **rock cores**, taking measurements of exposed rock layers, and photographing and making sketches to create a map of an area.

These scientists also spend time in the **LAB**. They use high-tech machines and do clever experiments on rocks to work out information like how old they are, what minerals they're made of, and how strong they are.

Geologists' work is really important - they are making discoveries all the time that increase our understanding of Earth!

WHO KNOWS WHAT THEY'LL DISCOVER NEXT?

AMAZING ROCKY PLACES

Rocks make up the whole world, so they're around us all the time, even if we don't see them. But in some places, rocks make up such a huge part of the landscape!

Monument Valley, USA

This huge area of red sandstone began to form about 300 million years ago! It has taken millions of years to wear into these shapes. The rock stacks here are called spires.

Mount Everest, Asia

On the border of Nepal and Tibet sits Mount Everest, the world's tallest mountain above sea level. Scientists discovered it's made of all three types of rock, and is getting taller every year!

Giant's Causeway, UK

By the sea in Northern Ireland, Giant's Causeway is made up of 40,000 columns of an igneous rock called basalt. It has been the inspiration of stories about giants crossing the sea to Scotland.

Painted Cliffs, Australia

The patterns on the Painted Cliffs in Tasmania, Australia, were created by iron-rich water which filtered through the rock and left patterns behind. The rock face has been shaped by ocean waves.

HOW TO FIND ROCKS

You don't have to be an expert to find rocks; you just need a few basic tools and to know what to look for!

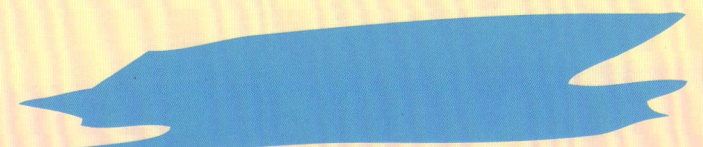 TO GET STARTED, YOU WILL NEED:

- A strong backpack for carrying your tools and finds
- Old towels and plastic bags to line your backpack
- Goggles to protect your eyes
- A geological rock hammer
- A chisel for splitting rocks
- A magnifying glass
- A camera for taking pictures of rocks that are too big to carry
- A notebook and pencil for writing down your findings

BEFORE YOU GO

Plan your trip carefully. Some places have rules about collecting rocks – make sure you know the rules where you're going so you can follow them.

STAYING SAFE

Always go with an adult and stick together. Pack food, water, and a mobile phone in case you need to call for help.

FINDING ROCKS

Search for rocks on the seashore, riverbank, and hillside. Turn them over as sometimes the interesting part is underneath. Sometimes, you may want to split rocks open to see inside – always ask an adult to do this.

DISPLAYING YOUR ROCKS

Clean your rocks with an old toothbrush and warm water. Rocks can be heavy, so use a strong shelf or box for your display. Use a piece of card to write the type of each rock and where it was found.

IGNEOUS ROCKS

The world of rocks really is fascinating! Here's some impressive igneous rocks.

DIORITE

- Minerals: Feldspar, mica, hornblende
- Formed: Deep underground from magma
- Features: Speckled with large, dark and light crystals

BASALT

- Minerals: Feldspar, augite, olivine
- Formed: In volcanoes from lava
- Features: Dark in colour, very **fine** grains of crystal, often full of tiny holes

ANDESITE

- Minerals: Feldspar, mica, hornblende
- Formed: In volcanoes from lava
- Features: Fine grains of grey crystals with some larger light and dark ones

OBSIDIAN

- Minerals: Quartz, feldspar, hornblende
- Formed: In volcanoes from lava
- Features: Like black glass, may have white snowflake-like crystals inside

DOLERITE

- Minerals: Feldspar, augite
- Formed: Deep underground from magma
- Features: Mostly dark with tiny speckles of light-coloured crystals

PEGMATITE

- Minerals: Quartz, feldspar, mica
- Formed: Deep underground from magma
- Features: Speckled with large white, pink, or black crystals

SEDIMENTARY ROCKS

There are so many cool rocks in the world! Here's some super sedimentary rocks.

CHALK

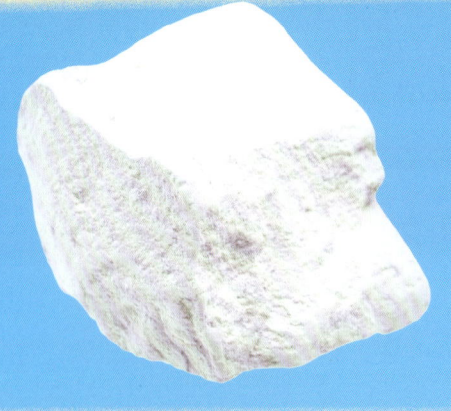

- Minerals: Calcite
- Formed: On seabeds
- Features: White and powdery, rubs off on your hands

SANDSTONE

- Minerals: Quartz, feldspar
- Formed: In shallow water, on seabeds, or riverbeds
- Features: Made of small grains of sand; smooth, sandy surface

SHALE

- Minerals: Clay, quartz, mica, feldspar
- Formed: In shallow water, on seabeds or riverbeds
- Features: Dark colour with some light specks, made of tiny grains

LIMESTONE

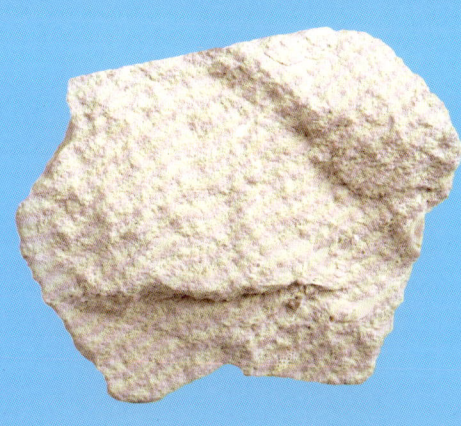

- Minerals: Calcite, dolomite, feldspar, quartz
- Formed: On the seafloor
- Features: Light-coloured stone with a grainy texture

COAL

- Minerals: None; mainly made of carbon
- Formed: In swampy forests
- Features: Black, with shiny patches; rubs off on your hands

CONGLOMERATE

- Minerals: Many minerals and pieces of rock (depending on where it is formed)
- Formed: On seabeds and riverbeds
- Features: Sandy grains, and chunks of different kinds of rock

METAMORPHIC ROCKS

As we have seen, rocks come in all shapes and sizes! Here's some amazing metamorphic rocks.

SLATE

- Minerals: Mica, chlorite, quartz
- Formed: Beneath mountains
- Features: Can be broken into thin layers, which may contain pyrite crystals

MARBLE

- Minerals: Calcite, clay, mica, quartz
- Formed: Deep inside Earth's crust
- Features: Patterned with stripes and swirls of colour

SERPENTINITE

- Minerals: Antigorite, chrysotile, lizardtile
- Formed: Deep inside Earth's crust
- Features: Bright red or dark green

HORNFELS

- Minerals: Cordierite, andalusite, garnet
- Formed: Near large areas of igneous rock
- Features: Dark and made of fine grains but with very sharp edges

SOAPSTONE

- Minerals: Talc
- Formed: Deep inside Earth's crust
- Features: **Microscopic** crystals make this rock very soft; can be scratched with fingernails

ECLOGITE

- Minerals: Pyroxene, garnet
- Formed: Deep inside Earth's crust
- Features: Mostly made up of red and green mineral crystals

RECORD-BREAKING ROCKS

All rocks have a story to tell, but these rocks are the most impressive of all.

How heavy a rock is depends on how **dense** the minerals inside it are. Peridotite and gabbro are two of the **DENSEST ROCKS** around.

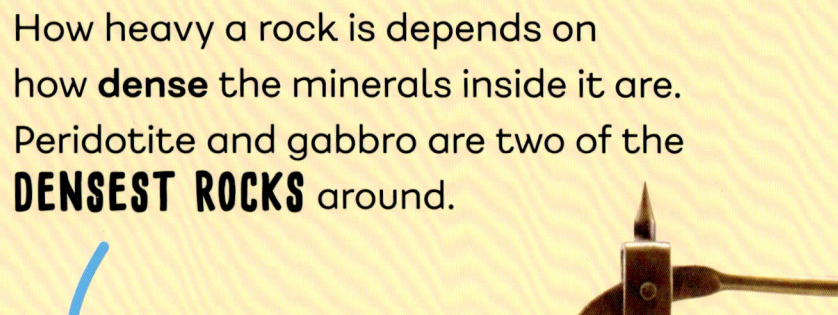

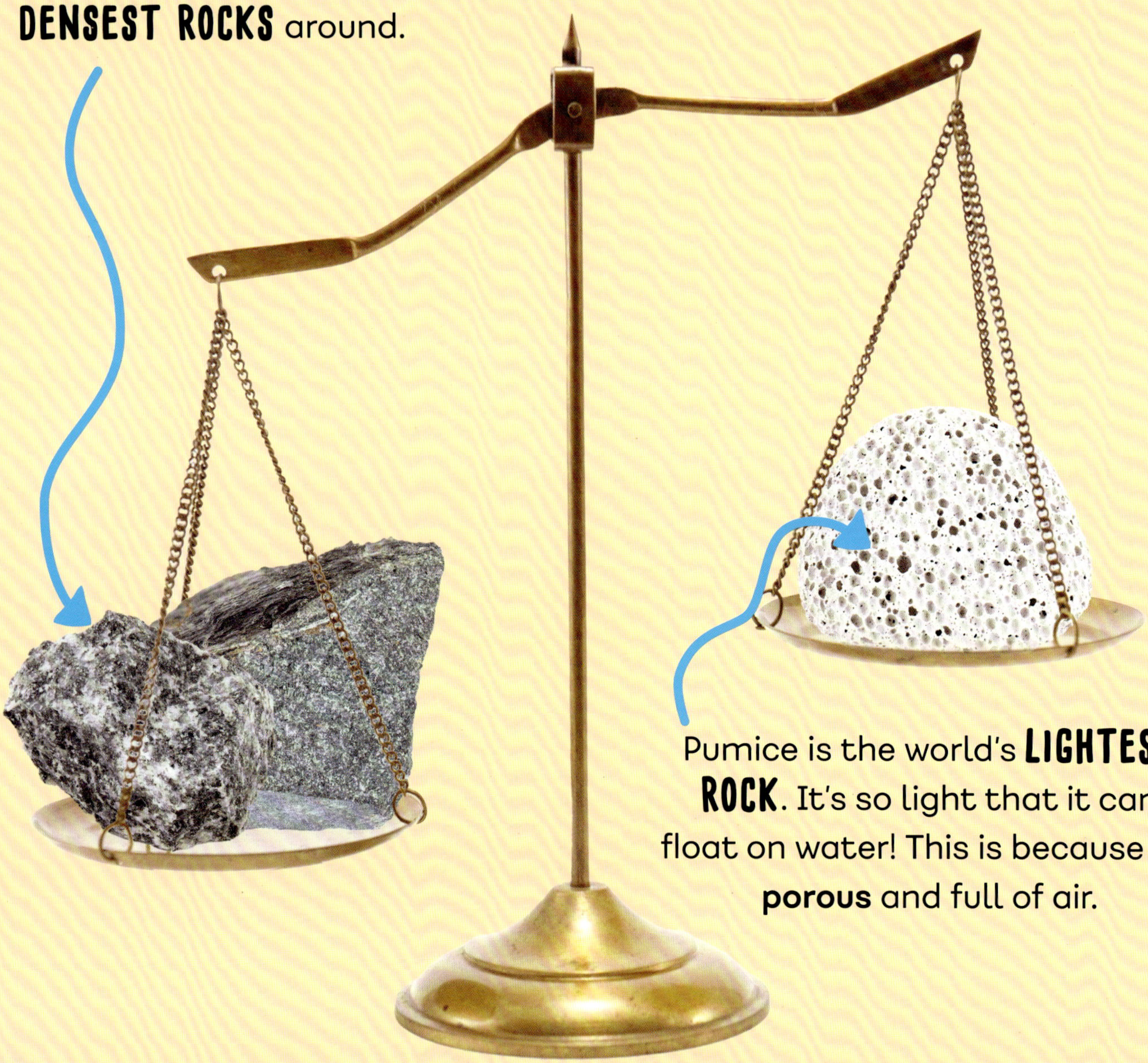

Pumice is the world's **LIGHTEST ROCK**. It's so light that it can float on water! This is because it's **porous** and full of air.

Thought to be 1.6 billion years old, Mount Augustus in Australia is the **BIGGEST ROCK** in the world! It's called Burringurrah by the local Wajarri community.

This rock is 715 metres (2,400 ft) tall.

The **OLDEST ROCK** on Earth isn't from Earth at all. A **meteorite** that landed in Australia contains tiny grains that are thought to be about 7 billion years old – that's older than the Sun!

The Richat Structure in Africa is made of igneous and sedimentary rock that have worn away over time to reveal **UNUSUAL** ring-shaped layers. It's often called the "Eye of the Sahara" because of how it looks from space!

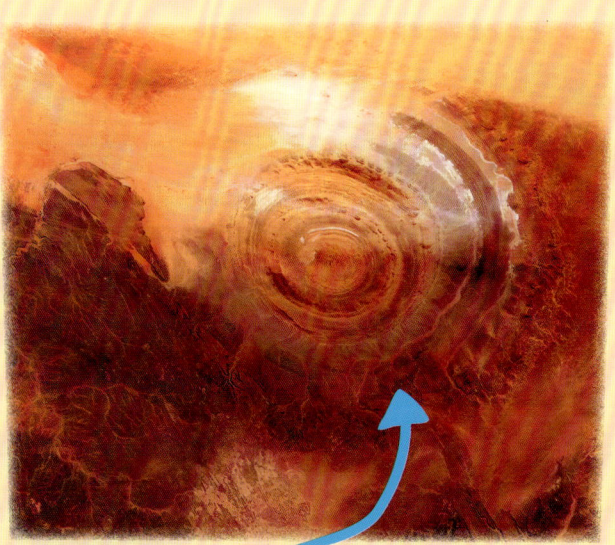

THE MOST UNUSUAL-LOOKING ROCK

TRUE OR FALSE?

There's lots to learn about rocks. How well do you know your rock facts?

Lots of rocks don't contain minerals!

FALSE! Obsidian, which is a glassy igneous rock, is one of the few rocks that contain no minerals.

Earth isn't the only planet made of rocks!

TRUE! Mars in our solar system, is often called "The Red Planet" after its red appearance. This is caused by high levels of iron within its rocks.

Rocks can tell us about past human lives!

TRUE! Thousands of years ago, ancient humans made carvings inside rocky caves which historians today use to learn about the past.

Sand is always yellow!

FALSE! The sand on the beaches of some volcanic islands, like Hawaii, is naturally black! This is because it's made of ground-up lava.

Basalt is the most common rock on Earth.

TRUE! It covers more than half of the whole planet! All ocean beds are made of this igneous rock.

31

GLOSSARY

Chisels – long metal tools that are used to chip away solid material, like rock or wood.

Climate – the long-term temperature patterns and weather conditions in an area.

Climate change – the change in climate (see above) over a long period of time.

Composition (rocks) – everything that makes up a material, including the amounts of each thing.

Crystals – a solid material that forms in a repeated pattern.

Dating (rocks) – to work out how old something is.

Dense – layers tightly compacted together.

Fine (grains) – incredibly small pieces of something.

Fossils – the remains or impression of plants and animals that lived long ago.

Grains – tiny, hard pieces of something. Rocks are made from mineral grains compacted together.

Meteorite – a rock that comes from outer space and falls to Earth's surface.

Microscope – a scientific tool that magnifies things so that they look much bigger.

Microscopic – something that is so small it can only be seen with a microscope.

Minerals – substances that are naturally found in things like rocks, sand, and soil. Many minerals form as crystals.

Molten – another word for melted.

Porous – something that has lots of holes. Air or liquid passes through porous materials easily.

Rock cores – long tubes of rock extracted from the ground that show the different layers inside.

Species – a group of living things that share characteristics and features.

INDEX

B
Basalt 19, 22, 31

C
Canadian Shield, Canada 13
Coyote Buttes, USA 11
Crystals 9, 22-23, 26-27

F
Fossils 11, 15

G
Gabbro 28
Geologists 16-17
Giant's Causeway, UK 19
Gneiss 12-13
Grand Canyon, USA 5
Granite 8-9, 12-13

I
Igneous rocks 6-7, 8-9, 13, 18-19, 22-23, 29, 30-31

Lava (see: Molten rock)
Limestone 11, 25

M
Magma (see: Molten rock)
Mars 30
Metamorphic rocks 6-7, 12-13, 18, 26-27
Meteorite 29
Minerals 7, 9, 10-11, 12-13, 17, 22-23, 24-25, 26-27, 28, 30
Molten rock 6-7, 8-9, 22-23
Monument Valley, USA 18
Mount Augustus, Australia 29
Mount Everest, Asia 18

O
Obsidian 30

P
Painted Cliffs, Australia 19

Peridotite 28
Pumice 28

R
Richat Structure, Africa 29
Rock formation 6-7, 8-9, 10-11, 12-13, 22-23, 24-25, 26-27
Rock hunting 20-21

S
Sandstone 10-11, 18, 24
Sedimentary rocks 6-7, 10-11, 18, 24-25, 29

V
Volcanoes 6, 8, 15, 22-23, 31

W
Wave Rock, Australia 5